Learn ✻ Color ✻ Align

Chakra Book

BALBOA.PRESS
A DIVISION OF HAY HOUSE

Copyright © 2024 144 Esoteric.

All rights reserved. No part of this book may be used or reproduced by any means, graphic, electronic, or mechanical, including photocopying, recording, taping or by any information storage retrieval system without the written permission of the author except in the case of brief quotations embodied in critical articles and reviews.

Balboa Press books may be ordered through booksellers or by contacting:

Balboa Press
A Division of Hay House
1663 Liberty Drive
Bloomington, IN 47403
www.balboapress.com
844-682-1282

Because of the dynamic nature of the Internet, any web addresses or links contained in this book may have changed since publication and may no longer be valid. The views expressed in this work are solely those of the author and do not necessarily reflect the views of the publisher, and the publisher hereby disclaims any responsibility for them.

The author of this book does not dispense medical advice or prescribe the use of any technique as a form of treatment for physical, emotional, or medical problems without the advice of a physician, either directly or indirectly. The intent of the author is only to offer information of a general nature to help you in your quest for emotional and spiritual well-being. In the event you use any of the information in this book for yourself, which is your constitutional right, the author and the publisher assume no responsibility for your actions.

Any people depicted in stock imagery provided by Getty Images are models,
and such images are being used for illustrative purposes only.
Certain stock imagery © Getty Images.

ISBN: 979-8-7652-5505-6 (sc)
ISBN: 979-8-7652-5504-9 (e)

Print information available on the last page.

Balboa Press rev. date: 08/23/2024

What is Chakra?

- 🌸 A chakra is like an energy wheel that moves the energy throughout our body.
- 🌸 There are seven main chakras in our body located between the base of our spine and the top of our head.
- 🌸 Balanced and open chakras are important for our physical, emotional and spiritual well-being.
- 🌸 Let us have fun learning about and coloring each chakra.

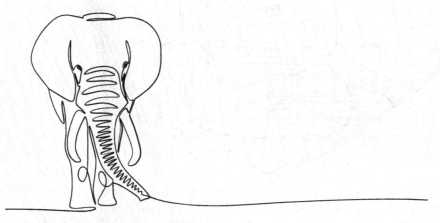

Tip: Saying and feeling the affirmation associated with each chakra while coloring, can help further balance and open our energy flow.

Self-awareness leads to a better life

1

Root Chakra

(Muladhara)

- Root Chakra, or Muladhara (means "root support" in Sanskrit), is located at the base of our spine.
- It is associated with survival, physical stability, grounding and more.
- Like roots of a tree, Root Chakra provides a foundation in our life.
- It provides us with the confidence to face any challenges in life.
- Affirmation: "I am"
  Sound: LAM
  Color: Red

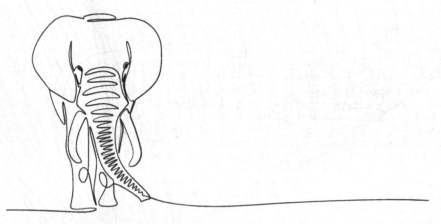

Tip: While coloring the Root Chakra symbol in shades of red, say and feel the affirmation below it.

I am safe, centered and grounded

2

Sacral Chakra

(Svadhisthana)

- Sacral Chakra, or Svadhisthana (means "dwelling place of Self" in Sanskrit), is located just below our belly button.
- This chakra controls our emotions and creativity.
- It supports overall joy in our life.
- Affirmation: "I feel"
  Sound: VAM
  Color: Orange

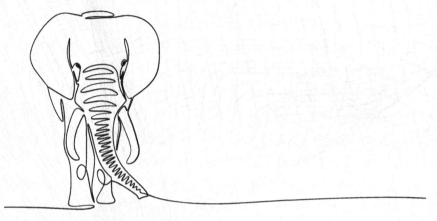

Tip: While coloring the Sacral Chakra symbol in shades of orange, say and feel the affirmation below it.

I am creative &
my life is full of joy!

Solar Plexus Chakra
(Manipura)

- Solar Plexus Chakra, or Manipura (means "city of jewels" in Sanskrit), is located right above the belly button.
- This chakra controls our power and confidence.
- It helps us be our true self and take responsibility for our life.
- Affirmation: "I can"
  Sound: RAM
  Color: Yellow

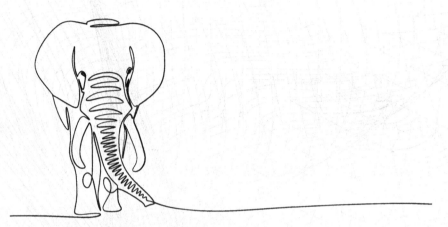

Tip: While coloring the Solar Plexus Chakra symbol in shades of yellow, say and feel the affirmation below it.

I am strong & confident

4

Heart Chakra

(Anahata)

- Heart Chakra, or Anahata (means "unhurt" in Sanskrit), is located in the center of our chest area.
- This chakra controls love, kindness and inner peace.
- It influences our ability to give and receive love.
- Affirmation: "I love"
  Sound: YAM
  Color: Green

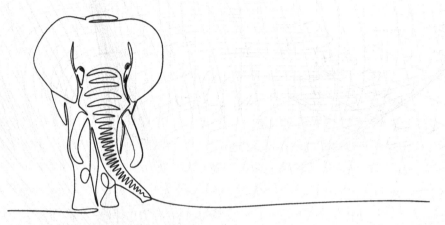

Tip: While coloring the Heart Chakra symbol in shades of green, say and feel the affirmation below it.

Throat Chakra
(Vishuddha)

- Throat Chakra, or Vishuddah (means "purification" in Sanskrit), is located in the center of our throat.
- This chakra controls our ability to speak our truth.
- It allows us to express ourselves truly and clearly.
- Affirmation: "I speak"
  Sound: HAM
  Color: Blue

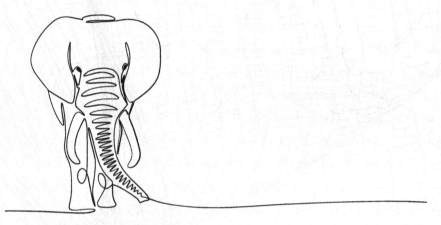

Tip: While coloring the Throat Chakra symbol in shades of blue, say and feel the affirmation below it.

I always speak
my truth

6

Third-Eye Chakra

(Ajna)

- Third-Eye Chakra, Ajna (means "to preceive" in Sanskrit), is located on our forehead, between the eyes.
- This chakra controls our imagination, intuition and wisdom.
- It allows us to see the big picture and align with our purpose in life.
- Affirmation: "I see"
  Sound: OM
  Color: Indigo

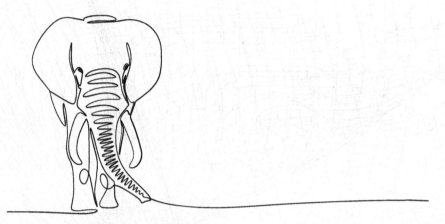

Tip: While coloring the Third-Eye Chakra symbol in shades of indigo, say and feel the affirmation below it.

I trust & follow my intuition

Crown Chakra

(Sahasrara)

- Crown Chakra, or Sahasrara (means "thousand-petaled" in Sanskrit) is located on top of our head.
- This chakra connects us to our creator/divine source.
- It allows us to feel clarity of thought and deeper connection to the universe.
- Affirmation: "I know"
  Sound: AUM
  Color: Voilet

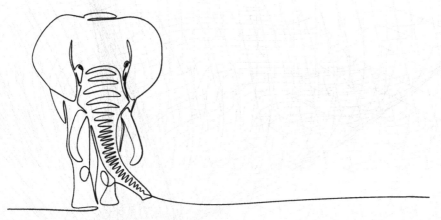

Tip: While coloring the Crown Chakra symbol in shades of voilet, say and feel the affirmation below it.

I am connected to
& one with all

Balanced & Aligned

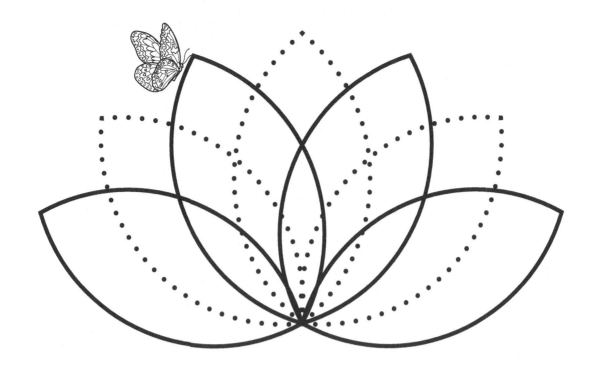

- Like any system in our body, balanced and aligned chakras are important for our overall physical and emotional well-being.
- When balanced, the life force energy flows smoothly throughout our body, balancing and aligning our inner and outer worlds.
- Chakra balancing can impact all areas of our life. Balance in life is something we create and it all starts within.

I am balancing my inner & outer worlds

Being in the flow

- ❁ A healthy lifestyle and practices like meditation, yoga, connecting with nature, positive thinking and affirmations can help keep our chakras open and our energy flowing freely.
- ❁ This makes us feel whole within ourselves and one with the universe.
- ❁ We become more receptive and open to attract more love, joy and abundance in our life.

Crown Chakra
Enlightenment
"I understand"

Location: Top of the head
Color: Violet

Throat Chakra
Intuition, awareness
"I see"

Location: Between the eyes
Color: Indigo

Throat Chakra
Communication
"I speak"

Location: Throat area
Color: Blue

Heart Chakra
Love, compassion
"I love"

Location: Center of the chest
Color: Green

Solar Chakra
Strength, confidence
"I do"

Location: Right above navel
Color: Yellow

Sacral Chakra
Creativity, emotion
"I feel"

Location: Right below navel
Color: Orange

Root Chakra
Stability, trust
"I am"

Location: Base of the spine
Color: Red

Printed in the United States
by Baker & Taylor Publisher Services